EARTH

SOLAR SYSTEM

Lynda Sorensen

The Rourke Corporation, Inc.
Vero Beach, Florida 32964

© 1993 The Rourke Corporation, Inc.

All rights reserved. No part of this book may be reproduced or utilized in any form or by any means, electronic or mechanical including photocopying, recording or by any information storage and retrieval system without permission in writing from the publisher.

Edited by Sandra A. Robinson

PHOTO CREDITS
All photos © Lynn M. Stone except cover and page 4, courtesy NASA; and page 21 courtesy USGS

Library of Congress Cataloging-in-Publication Data

Sorensen, Lynda, 1953-
   Earth / by Lynda Sorensen.
     p. cm. — (The Solar system)
   Includes index.
   Summary: Describes the physical nature, land sections, layers, and lifeforms of the Earth.
   ISBN 0-86593-275-1
   1. Earth—Juvenile literature. [1. Earth.] I. Title. II. Series: Solar system (Vero Beach, Fla.)
QB631.S67   1993   93-17007
525—dc20   CIP
                                                         AC

Printed in the USA

TABLE OF CONTENTS

The Earth	5
The Earth and Sun	6
A Special Planet	9
Life on Earth	11
The Earth's Land Surface	14
Water and Ice	16
The Earth's Crust	19
Beneath the Crust	20
Studying the Earth	22
Glossary	23
Index	24

THE EARTH

The Earth is a rocky, ball-shaped mass. It is one of the nine large heavenly bodies in our **solar system** called **planets.**

The Earth is the fifth largest of the planets. It is about 25,000 miles around.

The Earth's surface is made up of land, water and ice. The Earth is surrounded by invisible gases. They make up the **atmosphere,** which includes the air we breathe.

Earth's largest land sections—Asia, North America, South America, Europe, Australia, Africa and Antarctica—are called continents.

The ball-shaped Earth, seen from distant space

THE EARTH AND SUN

The Earth travels in a path, or **orbit,** around the sun once each year. The Earth's angle toward the sun changes during that time, so the amounts of sunlight and heat that reach the Earth change, too. These changes cause our seasons.

As the Earth travels around the sun, the Earth spins, or rotates. The part of the Earth facing the sun has daylight. The side away from the sun has night.

As the Earth rotates, the sun seems to "set"

A SPECIAL PLANET

The sun is 93 million miles from Earth. But enough of the sun's energy—its heat and light—reaches Earth to make life possible.

Plants grow by changing sunlight into food. Animals eat the plants, or they eat other animals that have eaten plants. In that way, the sun's energy flows through all life on Earth.

The Earth is a special planet. As far as we know, it is the only planet with life and large amounts of water.

Plants change sunlight into food

LIFE ON EARTH

Life on Earth exists in great variety. Thousands of kinds of plants and animals share forests, deserts, grasslands, **tundra** and bodies of water.

Some living things, or **organisms,** live in the air. Others live underground. Scientists discover "new" organisms every week.

The largest organisms, such as redwood trees and blue whales, are easily seen. However, many organisms are **microscopic**—they can be seen only through the powerful lens of a microscope.

The grasslands of East Africa support large numbers of plant eaters

The antelope eats the grass, and the lion eats the antelope

America's national bird seems to be screaming about pollution

Portage Public Library

THE EARTH'S LAND SURFACE

The Earth's land surface is like a great quilted blanket, made up of many different "pieces." There are valleys, canyons, mountains, plains, formations of rock and ice, and bodies of water.

Many of these features in the Earth's surface were created over thousands, even millions, of years. The Earth continues to change because of weather, ice, water, volcanoes and earthquakes.

The Rocky Mountains are the "backbone" of North America

WATER AND ICE

Three-quarters of the Earth's surface is covered by water. Most of the water is in the oceans. The fresh water we drink is in ponds, lakes, rivers and underground pools.

Some fresh water is in huge masses of ice. Much of the ice is found at both ends, or **poles,** of the Earth. **Glaciers** are great, frozen rivers. They are found in many mountain ranges.

At the south pole, Antarctica is covered by thick ice. Some of it is a mile deep!

The foot of Exit Glacier in Kenai Fjords National Park, Alaska

THE EARTH'S CRUST

The thick outer covering of the Earth is its rocky crust. In different places, the crust can be from about five to 25 miles deep.

Giant sections of the crust, called **plates,** sometimes shift and grind against each other. This causes earthquakes.

The Earth's crust is also shaken and shaped by volcanoes. Volcanoes are breaks in the crust through which gases and flaming liquid rock, called lava, explode.

Now "asleep," Mount Rainier in Washington was once a raging volcano

BENEATH THE CRUST

If you could peel the Earth like an onion, you would find layers of rock and other materials. Beneath the outermost peel, Earth's outer crust, is a layer of rock about 1,800 miles deep.

Beneath the rock, the inside of the Earth is so hot that it melts rock and metal. The lava that gushes from a volcano was heated inside the Earth.

There are layers of melted iron and nickel toward the Earth's center.

Flaming lava gushes from deep within the Earth

STUDYING THE EARTH

Many different kinds of scientists study planet Earth and its life. Biologists are the scientists who study organisms, for example.

Scientists know that many things people do damage air, water, land and life. Pollution, which is the release of harmful substances into our environment, is just one damaging activity.

One way that everyone can help our remarkable living planet stay healthy is to recycle. Save waste products for reuse.

Glossary

atmosphere (AT mus fear) — the air mass surrounding Earth

glacier (GLAY sher) — a huge, frozen river of ice and snow

microscopic (my kro SKAH pik) — able to be seen only through the powerful lens of a microscope; invisible without a microscope

orbit (OR bit) — the path that an object follows as it repeatedly travels around an object in space

organism (OR gan izm) — a living thing

planet (PLAN it) — any one of the nine large, ball-shaped heavenly bodies that orbit the sun

plates (PLATES) — the huge, moveable sections of the Earth's crust

poles (POLES) — the two exceptionally cold areas at opposite "ends" of the Earth; the north and south poles

solar system (SO ler SIS tim) — the sun, planets and other heavenly bodies that revolve around the sun

tundra (TUN druh) — the broad, treeless area of the Far North where plants grow low to the ground and soil thaws only a few inches deep each summer

INDEX

air   5, 22
animals   9, 11
Antarctica   16
atmosphere   5
continents   5
crust   19, 20
earthquakes   14, 19
energy   9
gases   5, 19
glaciers   16
heat   6, 20
ice   5, 14, 16
iron   20
land   5, 22
lava   19, 20
life   9, 11, 22
mountains   14
nickel   20

oceans   16
orbit   6
planet   5, 9
plants   9, 11
plates   19
poles   16
pollution   22
rock   14, 20
scientists   22
seasons   6
sun   6, 9
sunlight   6, 9
surface   5, 14
trees, redwood   11
volcanoes   14, 19
water   5, 9, 11, 14, 16, 22
whales, blue   11

Portage Public Library